Terrific Trucks!

FARM TRUCKS

I0808606

Ursula Pang

PowerKiDS press.

PK Beginners

Farms need trucks!

Tractors do a lot of farmwork.

JOHN DEERE
6195M

This tractor pulls a seeder.

This tractor lifts hay.

This bulldozer
moves dirt.

This sprayer truck waters plants.

This combine picks grain.

This grain truck takes the grain away.

This truck
moves animals!

This truck takes plants to stores.

Farms need a lot of trucks!

Published in 2026 by The Rosen Publishing Group, Inc.
2544 Clinton Street, Buffalo, NY 14224

Copyright © 2026 by The Rosen Publishing Group, Inc.

All rights reserved. No part of this book may be reproduced in any form without permission in writing from the publisher, except by a reviewer.

First Edition

Editor: Greg Roza
Book Design: Jeffrey Taylor

Photo Credits: Cover pilesasmiles/IStock.com; p. 3 AI Generated/Shutterstock.com; p. 5 ArliftAtoz2205/Shutterstock.com; p. 7 Bits And Splits/Shutterstock.com; p. 9 Bob Pool/Shutterstock.com; p. 11 Nick_Noppalit/Shutterstock.com; p. 13 AI Generated/Shutterstock.com; p. 15 Aleksandr Rybalko/Shutterstock.com; p. 17 mykhailo pavlenko/Shutterstock.com; p. 19 Gerard Koudenburg/Shutterstock.com; p. 21 AI generated image/Shutterstock.com; p. 23 Bilanol/Shutterstock.com.

Library of Congress Cataloging-in-Publication Data

Names: Pang, Ursula author
Title: Farm trucks / Ursula Pang.
Description: [Buffalo] : PowerKids Press, [2026] | Series: Terrific trucks! | Audience term: juvenile
Identifiers: LCCN 2024053752 (print) | LCCN 2024053753 (ebook) | ISBN 9781499451658 library binding | ISBN 9781499451641 paperback | ISBN 9781499451665 ebook
Subjects: LCSH: Farm trucks--Juvenile literature
Classification: LCC S711 .P24 2026 (print) | LCC S711 (ebook) | DDC 631.3/72--dc23/eng/20250131
LC record available at https://lccn.loc.gov/2024053752
LC ebook record available at https://lccn.loc.gov/2024053753

Manufactured in the United States of America

Some of the images in this book illustrate individuals who are models. The depictions do not imply actual situations or events.

CPSIA Compliance Information: Batch #CSPK26. For further information contact Rosen Publishing at 1-800-237-9932.

Find us on